IMAGES
of America

TRAVIS
AIR FORCE BASE

To all brave souls who stood for this country
and risked their lives for its preservation . . . I am in awe.

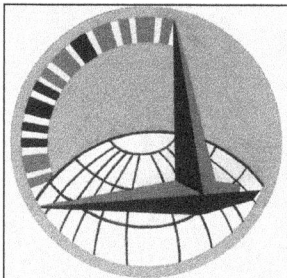

AIR TRANSPORT COMMAND

STRATEGIC AIR COMMAND

U.S. AIR FORCE MILITARY TRANSPORT SERVICE

MILITARY AIRLIFT COMMAND

AIR MOBILITY COMMAND

AIR FORCE RESERVE COMMAND

AIR MOBILITY COMMAND

IMAGES
of America

TRAVIS
AIR FORCE BASE

Diana Stuart Newlin

ARCADIA
PUBLISHING

Published by Arcadia Publishing
Charleston, South Carolina

Library of Congress Catalog Card Number: 2004111077

For all general information contact Arcadia Publishing at:
Telephone 843-853-2070
Fax 843-853-0044
E-mail sales@arcadiapublishing.com
For customer service and orders:
Toll-Free 1-888-313-2665

Visit us on the Internet at www.arcadiapublishing.com

A C-5 Galaxy flies over San Francisco in the 1980s.

ON THE COVER: RB-36s (reconnaissance B-36s) fly over Travis AFB around 1955. The B-36 named the Peacemaker was a unique aircraft in that it had six reciprocating "pusher" engines located at the trailing edge of the wing, along with four jet engines that were pod mounted. The B-36 squadrons stationed at Travis AFB, with their capability of delivering nuclear weapons, played a vital part in maintaining the peace throughout the 1950s and during the early stages of the cold war.

CONTENTS

Acknowledgments

When I was assigned as the deputy curator at the Travis Air Museum in 1999, I never imagined how this assignment would change my life and alter the way I saw things. I never knew there were so many heroes out there who stood the course, fought the good battle, and shielded their families from their violent memories. They never said a word—didn't want their families to ever worry about the things that went bump in the night. And I don't think they particularly wanted to relive the events either. They fought in defense of the freedom that their country and their loved ones enjoy—they felt no further comment was necessary. I do hope their families and others appreciate what these people went through. You don't go to war and come home with clean clothes.

I have met many veterans at the museum who honestly shared countless details about their experiences that still bring me to tears. It's never old hat and never the same old story. What these veterans can't express in words, they often express with their eyes and with their silence.

This book is a pictorial history of some of the lives and times at Travis Air Force Base, California, and is connected with much of America's history since 1942. My desire is not only to honor Travis personnel, but also to honor every soul who touched its soil either deploying or returning from a world many will never know. I know Travis is special to you too.

Because it is not possible to describe all the details of every mission and operation, this book serves as an overview. After searching through thousands of old, unmarked pictures and digging deep into hazy memories, I do not claim this work serves as a technical manual. It does, however, strive to rekindle a few memories and proud moments for those who lived them, and to share their experience so others might appreciate them.

I wish to thank Dr. Gary Leiser for his historical contributions to this book and his guidance over the museum, Senior Master Sgt. Ben Reed (ret.) for his dedication and creation of a superb archival library, Master Sgt. Joe Inocencio, and all the hard-working volunteers, who for years continue to show up every week bright and early. I can never thank you enough.

Most of the information in this book comes from the following sources: *A History of Travis Air Force Base* by Dr. Gary Leiser, and the *Tailwind* newspaper, Travis Air Force Base, California. For more information on the Travis Air Museum*, please visit the museum website at www. jimmydoolittlemuseum.org

*Renamed the Jimmy Doolittle Air & Space Museum in 2000—special thanks to the Doolittle family for this honor.

One

ORIGIN OF THE BASE

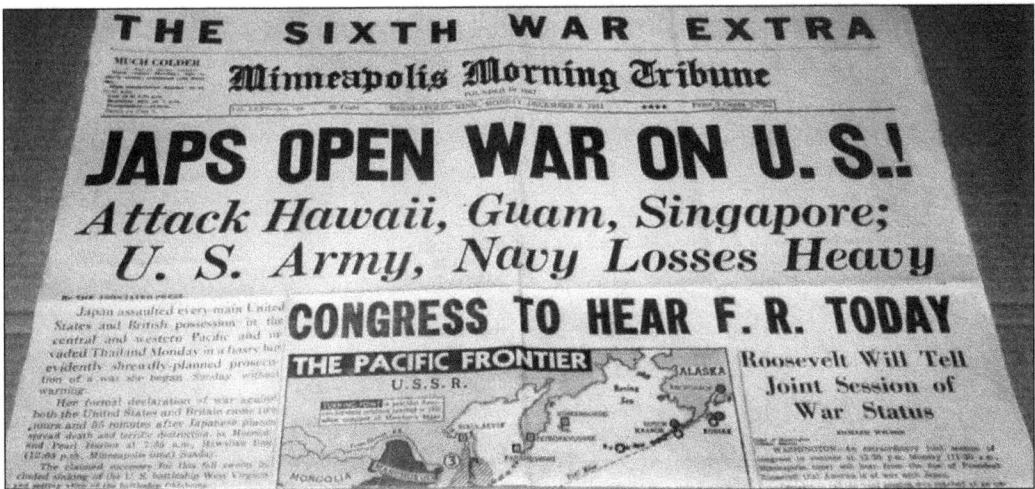

It was December 7, 1941—a date that would live in infamy, as President Franklin D. Roosevelt foretold after Japan targeted a surprise attack on Pearl Harbor, Hawaii. This aggression stunned and staggered the military as it occurred in the wake of peace negotiations, and left the United States reeling and seeking a counter attack as the Japanese military continued to defeat naval forces throughout the Pacific region.

Meanwhile, across the Pacific Ocean, the future of Travis Air Force Base in Fairfield, California, was emerging. The West Coast desperately needed a strategic defense against a Japanese invasion, and construction of a military installation was planned following the attacks. As it was located near the cities of Fairfield and Suisun, it was originally named the Fairfield-Suisun Army Air Field. It would later be renamed Travis Air Force Base.

Only months later, on April 22, 1942, the Office of the Chief of Engineers in Washington, D.C., authorized spending $998,000 for the construction of two runways and a few temporary buildings on a 945-acre site located six miles east of these twin farming communities. The wind currents there were powerful, making it an excellent location for aggressive pilot training. The government purchased land from local ranchers at an average cost of $50 per acre. On July 6, bulldozers went to work and by September runways and operations buildings were completed.

On October 13, 1942, the War Department assigned the newly planned facility, designated the Fairfield-Suisun Army Air Base, to the Air Transport Command (ATC).

These two photos show Station 10, Pacific Wing, Fairfield-Suisun Army Airfield in Fairfield, California. Air Transport Command took charge of the base on February 8, 1943, and activated it on May 17.

Two

WORLD WAR II

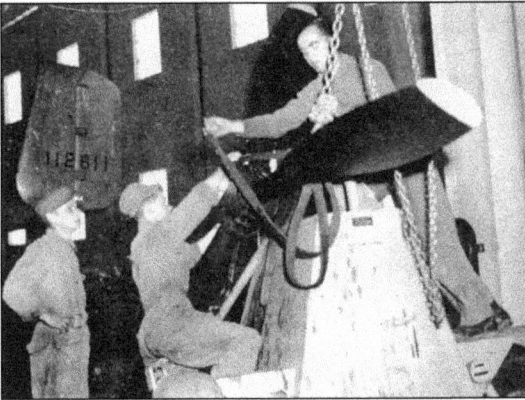

Between 1943 and 1945, the airfield's primary function was to prepare aircraft and their crews for the war in the Pacific. Thus it became the major jumping-off point on the West Coast for thousands of aircraft, chiefly bombers and transports, but also a few fighters. The bombers and transports included the B-24, B-25, B-17, B-29, AT-17, A-20, A-26, C-47 and C-54. Seen here are airmen uncrating a B-24 propeller at Station 10; fueling a transient—an AT-17/the Bamboo Bomber from the Pacific Fleet—and a line mechanic working on a B-25. Today the A-26, B-29, and C-54 are all part of the Travis Air Museum's aircraft collection.

During World War II, the Army Air Force significantly increased the size of the base.

This grounds maintenance crew takes a break from their beautification detail during the 1940s.

The military lifestyle developed.

The Service Club was the "army parlor" for GIs and WACs (Women's Army Corps). It had a floor for dancing, giving all a chance to relax and forget about waging war for a few stolen moments.

Enlisted men dating WACs could call at the WAC Squadron dayroom. Serving in a "liaison" capacity between the dayroom and the enlisted women's barracks was a WAC runner who was on duty during the evening hours. Monday through Friday, a midnight bed check was enforced in the WAC Squadron. Curfew on Saturday night was 0130, and on Sunday night 0100. Escorts were expected to return their dates to the WAC area at least 15 minutes before bed check!

Times were hopping. Only 18 months after it opened, the base was transformed into one of the nation's largest and most important military air trans-shipment points as witnessed in these flight line views of B-24s.

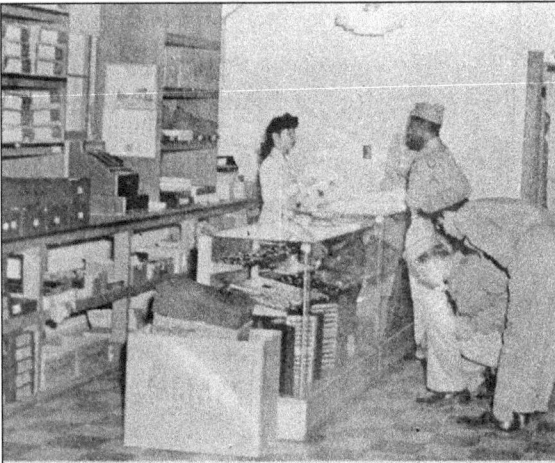

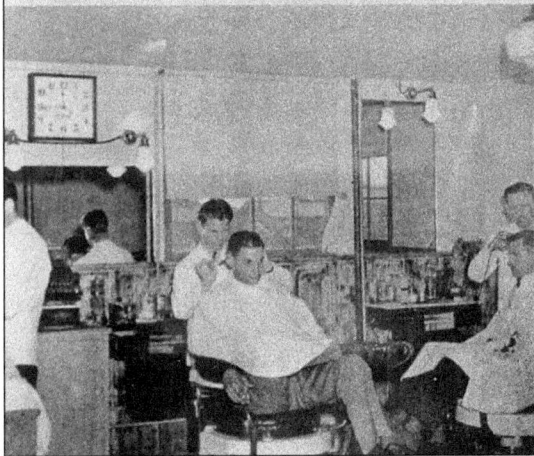

In early 1943, the airfield added 1,312 acres, and on June 30, 1945, another 1,145. As the size increased, base amenities multiplied. Subsequent expansion in the 1950s and 1960s increased the area to more than 6,000 acres.

The final assault on Japanese forces came in 1945 as U.S. crews prepared their flight plans and ATC shifted approximately 95 C-54s from its North Atlantic and North African services and from the Ferrying Division to its Pacific routes. All systems were go.

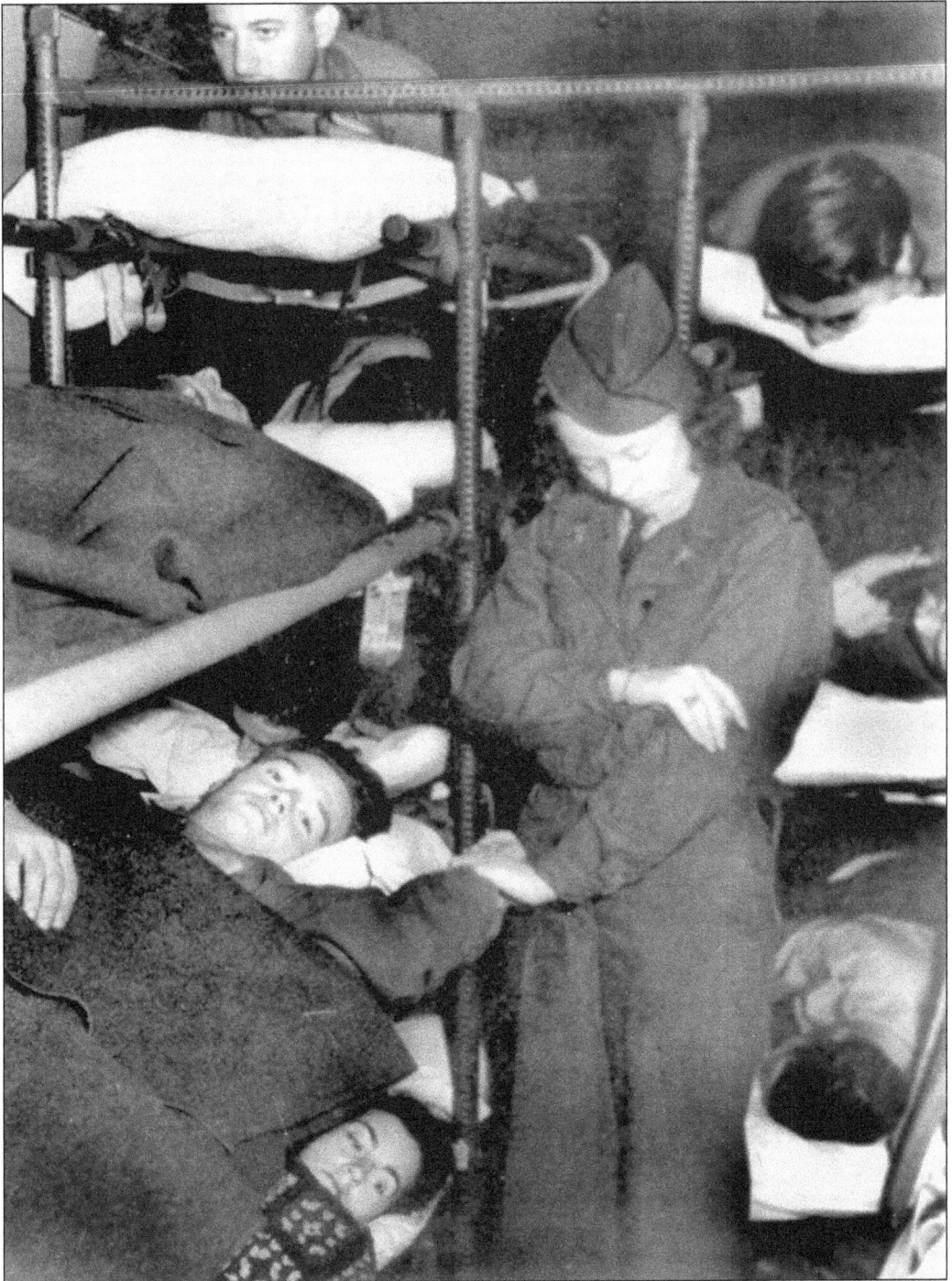

At its peak, ATC operations conducted "Mission 75," which included nearly 250 C-54s in the Pacific that flew occupation forces to Japan and returned wounded veterans and liberated POWs to the United States.

One of the most historic missions of that era was a flight from Fairfield-Suisun on August 10, 1945, en route to Ie Shima (near Okinawa) to carry Japanese envoys to the surrender conference in Manila, following the August 15 Japanese surrender. Peace at last!

On September 18, 1947, President Harry Truman officially created the United States Air Force as a separate military service. Consequently, Fairfield-Suisun Army Air Base became Fairfield-Suisun Air Force Base. Despite some lingering confusion in terminology between the terms "field" and "base," the latter word was adopted as the uniform Air Force designation in January 1948. Heretofore the two terms had been used more or less interchangeably, so that the Fairfield-Suisun installation was called both an Army Air Base and an Army Air Field.

A 1948 group of NCOs are seen after taking the test for warrant officer.

In 1948, the Defense Department combined the Air Force's Air Transport Command and the Navy's Air Transport Service to form the Military Air Transport Service (MATS). Air Force and Navy personnel congregate outside the headquarters building.

Three

KOREAN WAR

As the largest West Coast air terminal and an ATC base, Fairfield-Suisun came under the jurisdiction of MATS until May 1, 1949, when the Strategic Air Command (SAC) assumed control. Pictured above is a C-124 Globemaster ("Old Shakey") performing medevac (medical evacuation) duties. The outbreak of the Korean War in 1950 strengthened the importance of the base as one of SAC's main West Coast facilities and changed the operation's tempo.

The 9th Bombardment Wing immediately increased its local training sorties and practice bombing runs in preparation for deployment to Korea. It was during a deployment to Korea that Gen. Robert F. Travis was killed when his B-29 crashed near the end of the runway during a nighttime take-off. An explosion also resulted when the bomb bay impacted. Sixteen personnel perished.

Due to Travis's popularity and his outstanding military record, following the accident the base was renamed Travis Air Force Base on October 2, 1950, and a ceremony finalized it on April 20, 1951. Honored guests and officials at Travis Air Force Base watched officers and enlisted men pass in review to honor the late Brigadier General Robert F. Travis. Shown here, from left to right, are (front row) Mrs. Robert F. Travis, widow of the late base commander; Honorable Earl Warren, Governor of California; Brig. Gen. Joe W. Kelly, 14th Air Division Commander; Maj. Gen. (retired) and Mrs. Robert J. Travis, parents; and Maj. Gen. Charles I. Carpenter, chief of Air Force Chaplains; (back row) Maj. Gen. Emmett O'Donnell, 15th Air Force commanding general; Miss Jayne Travis, daughter; Lt. Col. William J. Travis (Robert's brother); Mrs. Edmund A. White (Robert's sister); and Col. C.J. Cochrane, former deputy commanding officer.

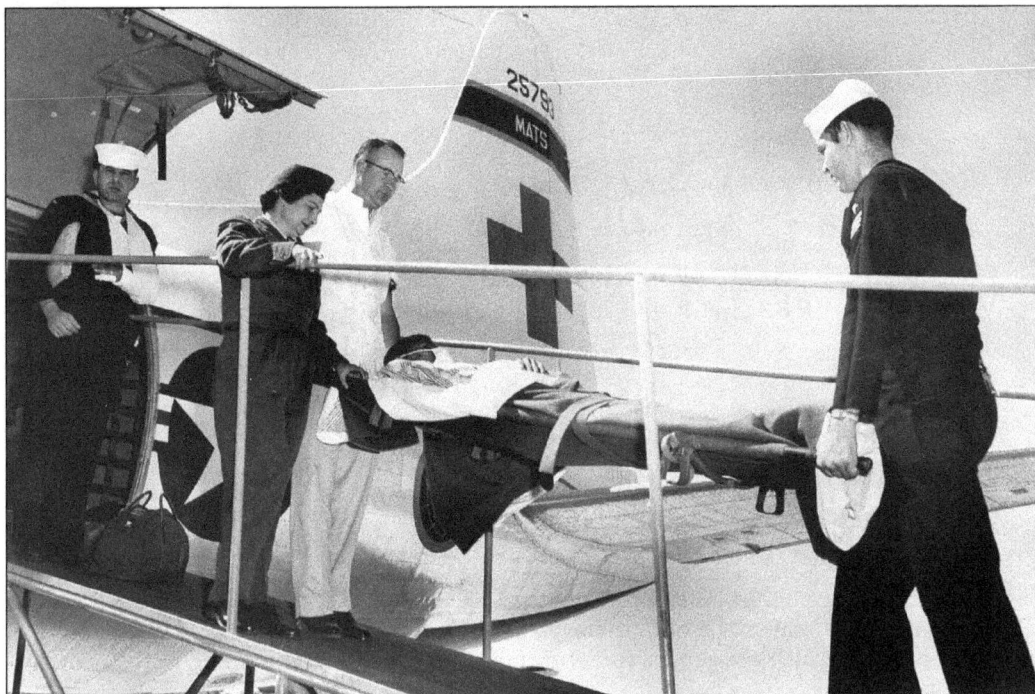

During the early stages of the Korean War, evacuation by air became the preferred method of returning wounded soldiers to hospitals. Pictured here are the many wounded deplaning a C-54, bearing witness to the rapid increase in American casualties.

A C-54 takes off daily with new patients headed stateside, following the policy of transferring the injured to domestic aeromedical flights.

The large, new "hospital on the hill" at Travis, which had opened shortly before the outbreak of the Korean War, became the central receiving point for casualties transported to the mainland by domestic airliners.

Another C-54 perseveres on medevac duty in this photograph. Between 1950 and 1953, the number of patients arriving from overseas each month averaged more than 2,000.

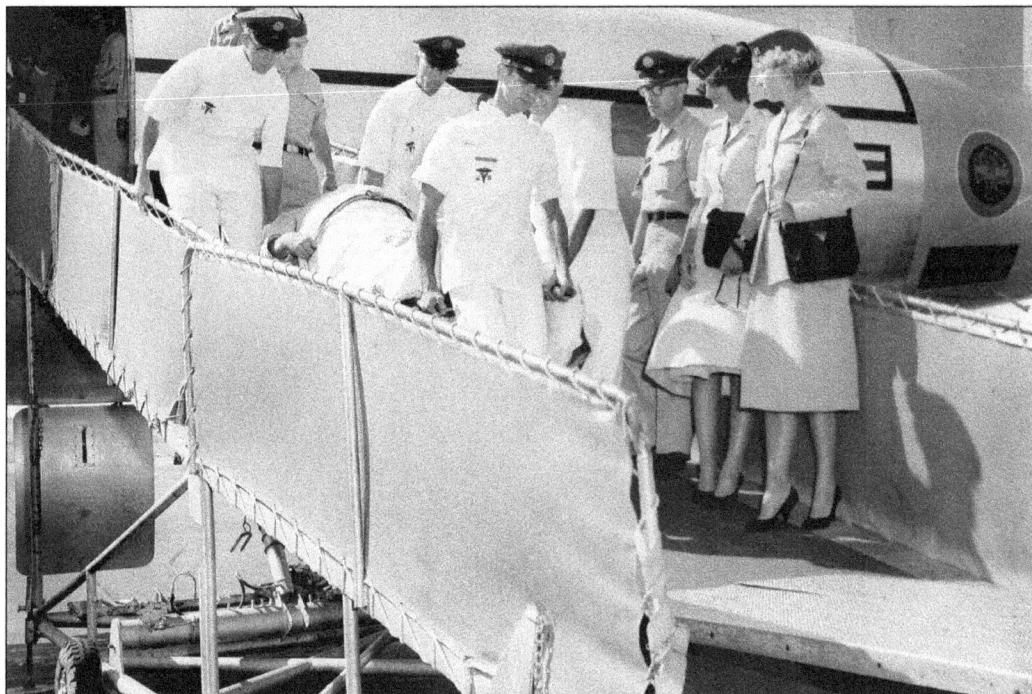

Travis's hospital was staffed with 26 officers, 10 nurses, 72 enlisted personnel, and 42 civilians, but this limited supply of personnel was greatly challenged to give adequate care to the high number of casualties.

The hospital staff, which was racially integrated, included several African-American doctors. This was unusual in 1950, when most American military units were still segregated.

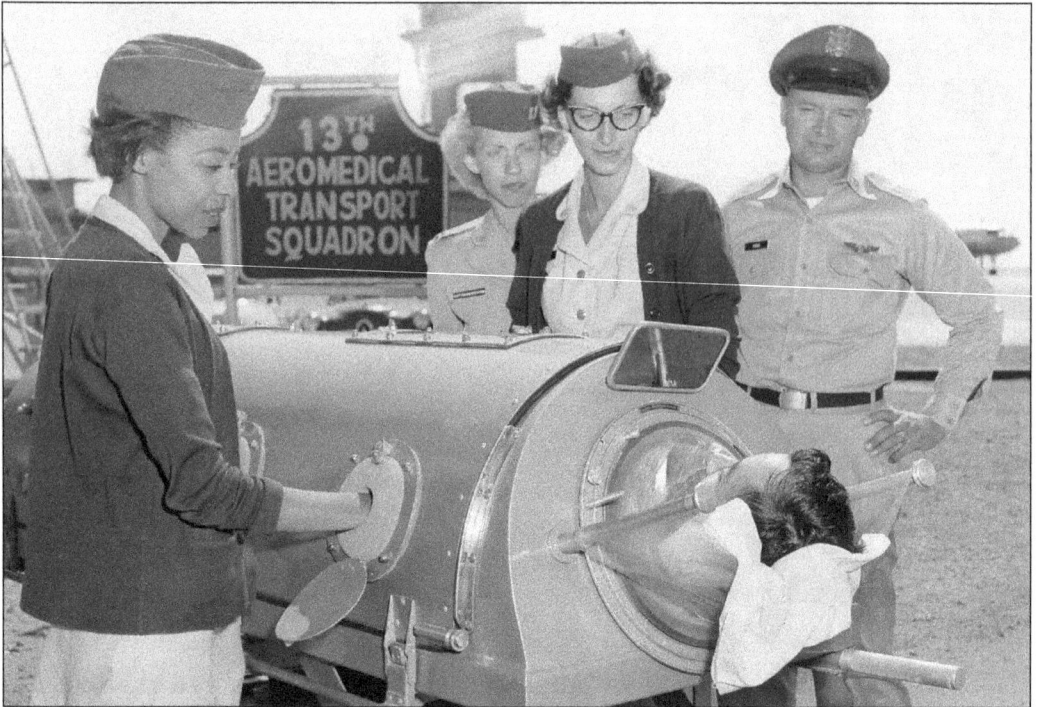

Due to the overwhelming flood of American casualties, the small staff had to develop new methods and procedures in order to handle the many wounded soldiers in a quick and orderly fashion.

This was no small task, as by early 1951 many of the wounded were arriving at the base within 48 hours of receiving combat injuries.

As the planes touched down, many casualties were still reeling from their experiences, often still shell-shocked, as they had only departed from the combat zone hours earlier.

Although the hospital was located right next to the runway, the volume of casualties still overwhelmed the facility.

As the C-124 airplanes circled the runway daily. many eyes of the wounded were at the round windows to catch the first glimpse of the California sunshine beaming down on Travis AFB—Travis AFB, U.S.A! One soldier was quoted as saying, "Well I made it. All the way back. All the way back."

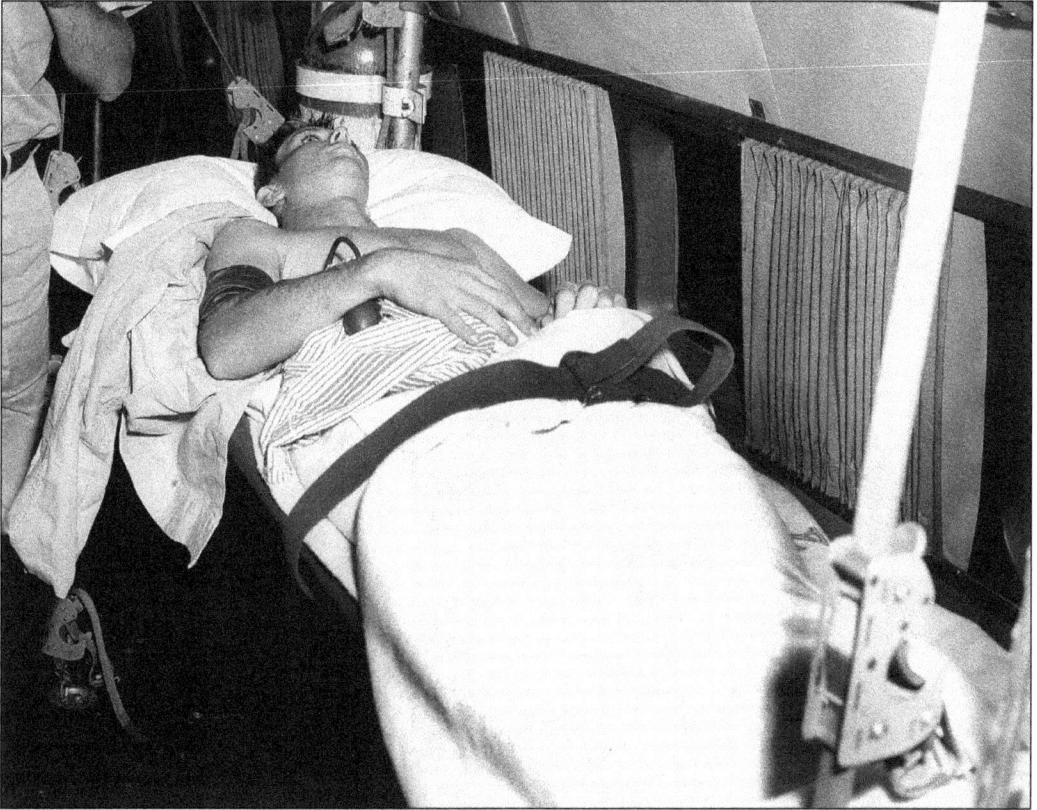

And yet, coming directly from the combat zone to the safety of Travis AFB was often a difficult transition for wounded soldiers. On one specific occasion as the plane approached the runway, for the first time since departing the war zone, "the worst psycho patient on board relaxed from his incoherent thrashing and no longer tugged at his restraints." His recent history included too much combat climaxing with a live shell bursting near him, which sent him over the edge. The doctor had said he was unaware of his surroundings, but on touchdown somehow he knew *he was home.*

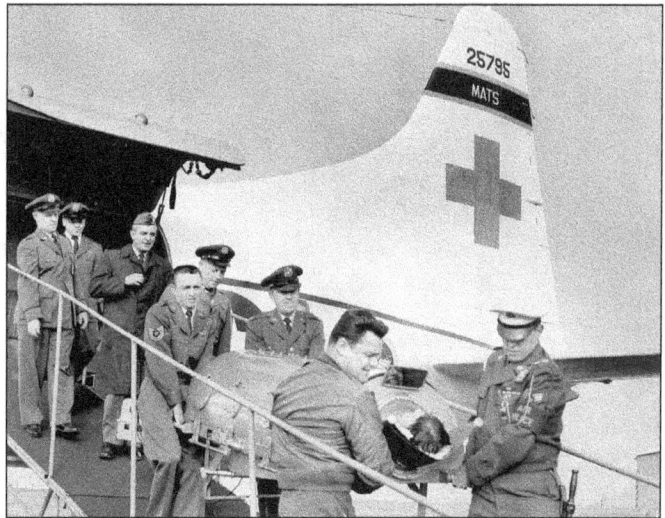

Another soldier was quoted, "I thought I'd be sore about the first look at the States because they'd forgotten about the dead and wounded in Korea, but I've never seen anything so marvelous in all my life. I love it. It's wonderful. It's not Korea or any place else. It's America."

BOEING B-29 SUPERFORTRESS
'COMMAND DECISION'
WORLDS MOST DECORATED BOMBER

Strategically, from the beginning of the Korean War, the Far East Air Forces had a continual need for additional combat aircraft. The transport of aircraft from the United States by ship was slow, so long-range ferry flights were organized. In order to avoid small aircraft losses over the Pacific, B-29 aircraft from Travis flew weather reconnaissance, and escorted navigation aircraft for the movement of smaller aircraft such as B-26s and F-84Gs, en route to combat zones in Korea.

Pictured here is a Travis B-29 crew.

The Special Services Office and the Hollywood Coordinating Committee made arrangements for celebrities to visit the base to entertain the wounded and deployed troops. Pictured at top is actress Elizabeth Taylor. Below, world heavyweight champ Joe Lewis greets troops. (USAF photographs by Tom Pierce.)

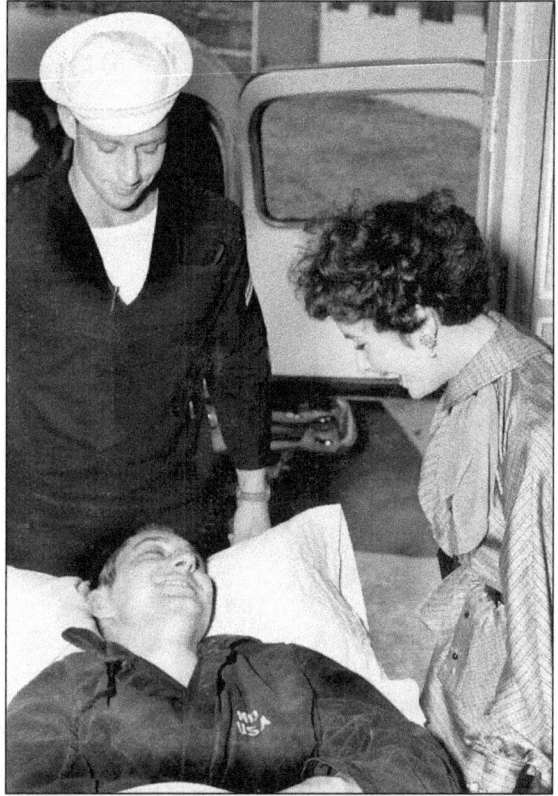

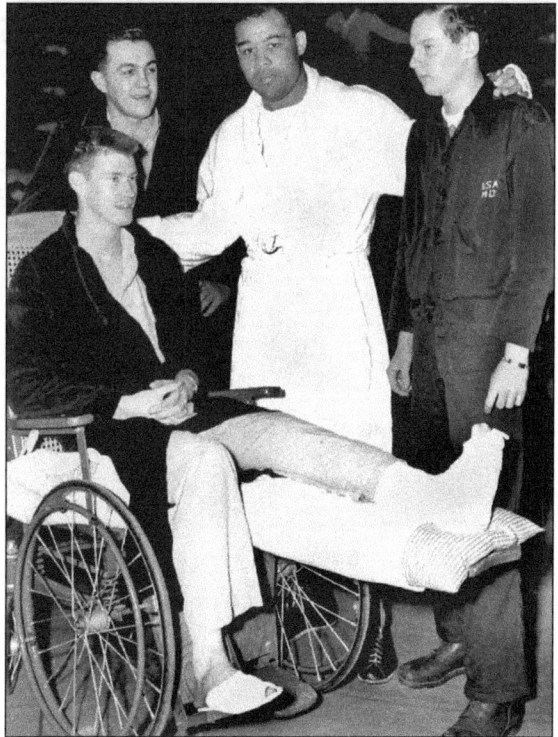

Many personalities performed tirelessly in support of their country. One serviceman said, "Whenever possible I used to watch the performances. Keenan Wynn seemed to be the group leader. I remember on Christmas Eve 1950, after the main performance, Debbie Reynolds and another pretty blonde lady stayed on and sang to the troops, anything they wanted until 2 a.m. Pretty nice, I thought of those two young ladies." Pictured at left are Debbie Reynolds and Carlton Carpenter and, according to the photo's donor, the image below shows Donna Reed giving a musical performance. (USAF photographs by Tom Pierce.)

This practice became so popular it was named Operation Starlift. Helmeted, rifle-carrying soldiers, marines, sailors, and airmen relaxed for a few moments and momentarily forgot that they would soon be on their way to combat. Pictured below is performer Danny Kaye. (USAF photograph below by Tom Pierce.)

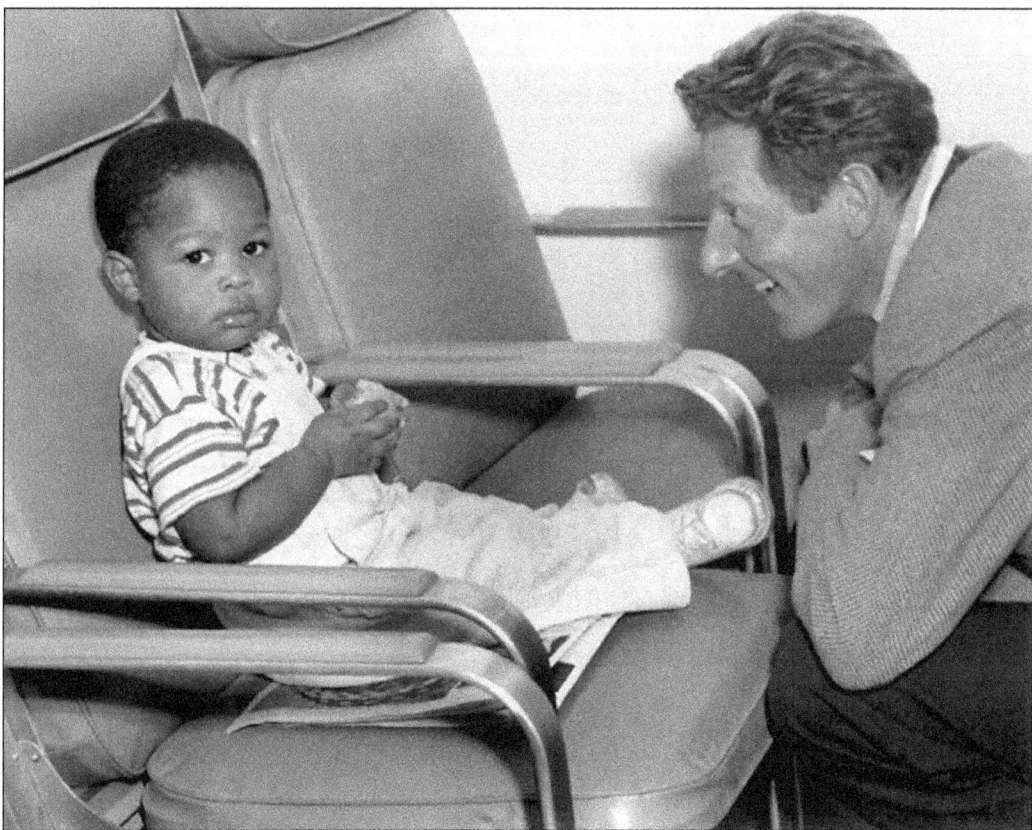

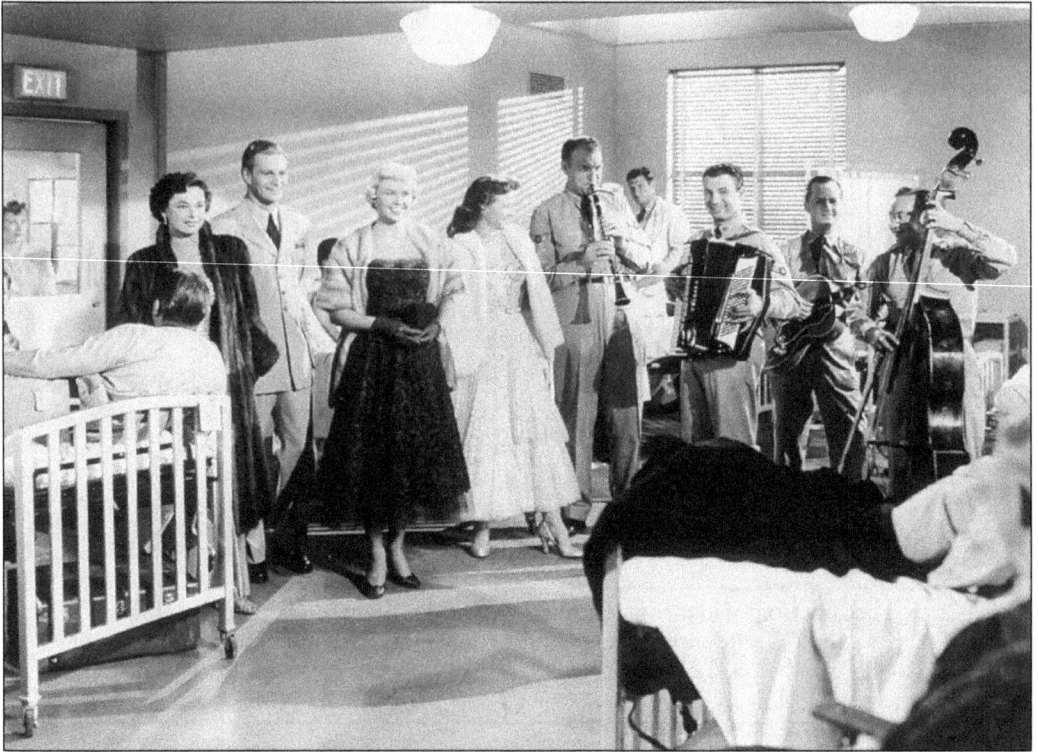

Doris Day shines at Travis in the above photo. The Warner Brothers motion picture *Starlift* was made at Travis and based on Travis's Operation Starlift. Said one airman, "The movie *Starlift* was filmed while I was at Travis. The movie was basically a movie about stars coming to the base hospital and the terminal at Travis to entertain the troops coming from and going to Korea. I saw the movie on base and then again at home at the Falls Theatre in Chippewa Falls, Wisconsin."

Though the stars shine bright for airmen, back in the hospital, in traditional fashion, the military nurses have always been at the heart of all battles and their contributions can never receive enough applause. These true angels of mercy brought the most welcome light to the wounded during the darkest time in their lives.

Troops anxiously await their homecoming in Korea.

And finally there was peace! Now they were almost home and flying over San Francisco at last! The Korean War lasted until 1953 when an armistice was signed which left a stalemate. The United States accepted the reality of two Koreas—a country divided at the 38th parallel.

Four

NEW AIRCRAFT AND
ORGANIZATIONAL CHANGES

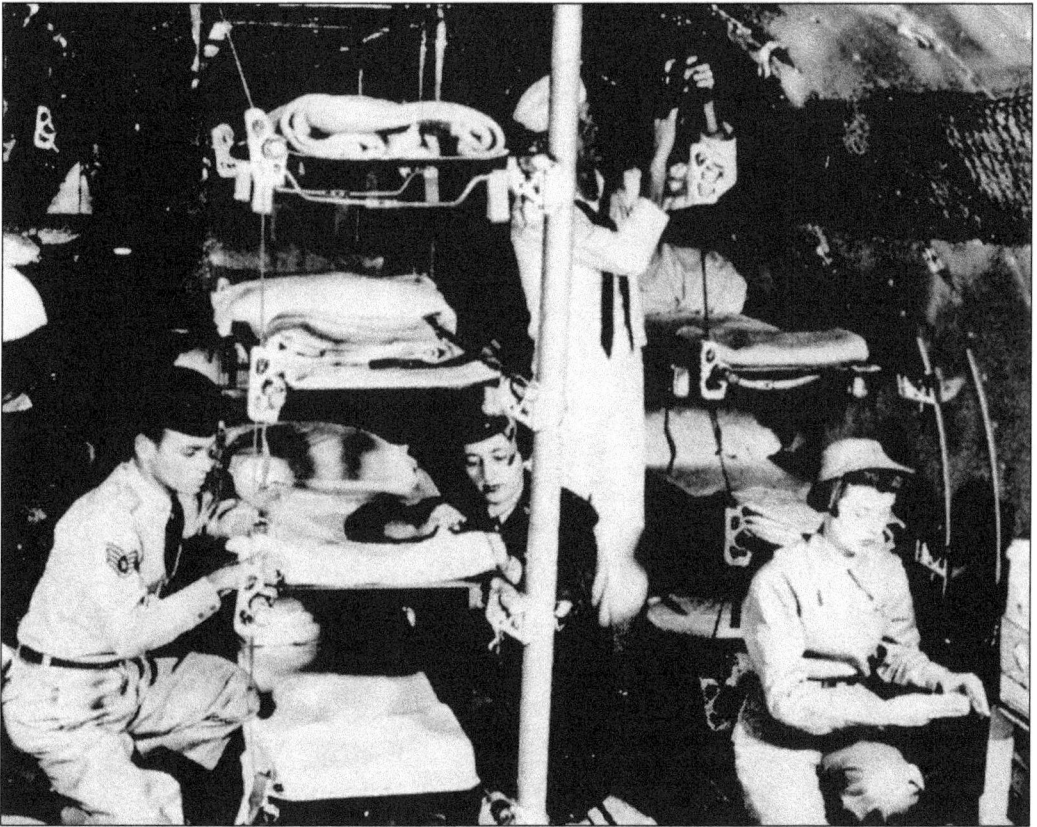

The Western Transport Air Force was assigned to Travis as part of the 1958 reorganization of MAT's subordinate commands. WESTAF's mission combined wartime readiness training, logistical airlift operations, aeromedical evacuation, and peacetime cargo airlift. Pictured above is a C-97 staffed for action. WESTAF consisted of one air division, one Naval air transport wing, four Air Force transport wings, two Air Force heavy troop carrier wings, an aeromedical transport group, and an air transport group.

In 1959, Travis AFB welcomed its first B-52 Stratofortress. It was introduced as part of SAC's expansion program linked to their new operational philosophy known as "satellite dispersal." Signifying a new era for Travis, the jet age had arrived—the beginning of the end for propeller planes.

It was out with the old as the prop plane era slowly came to a close. Pictured, from left to right, are the C-124 "Old Shakey," the C-121 Constellation, and the C-97 Stratofreighter. In 1960 as the jet age dawned, the old long-distance propeller transports were still flying under the 1501th Air Transport Wing. The C-97 Stratofreighter—the oldest strategic air transport—was built in 1947. The last two C-97 squadrons, the 47th and the 55th, were inactivated in March 1960. The aircraft were retired from active duty and reassigned to the Air National Guard.

A C-133 is viewed at an Open House with a C-124 in the background, but its days are numbered. The C-124 Globemaster (1953), was a favorite and had an excellent safety record. However in 1958 one crashed into the runway killing six Air Force personnel—the worst accident since General Travis's B-29 crash. Still, its popularity soared and its number increased to 37 in the fleet until the newer C-133 began to take over in 1958.

The C-133 Cargomaster, the most modern of the propeller aircraft, earned a new mission. Its 12-foot "clam shell" petal doors allowed it to transport the newly developed intercontinental ballistic missiles (ICBM).

A Travis C-135 stands proudly with her 44th Air Transport squadron crew after breaking four world records by lifting 66,000 pounds of cargo to 47,171 feet. The introduction of new jet-age technology into the MATS fleet between 1961 and 1965 led to a true airlift revolution and three new types of airlifters were assigned to Travis. The first was the C-135B Stratolifter, which served as the first jet medical evacuation flight, rushing as many as 38 critically ill patients from Asia to Travis in just 9 hours and 7 minutes.

On April 4, 1963, the second airlifter—the C-130 Hercules—arrived, activating the 86th Air Transport Squadron and the 1513th Organizational Maintenance Squadron. The activation of the 86th brought a total of six flying squadrons to Travis.

A Travis C-124 approaches the runway south of Tachikawa, Japan. All MATS/MAC aircraft flying from Travis were assigned to the 1501st Air Transport Wing, including this C-124. Other aircraft flown from 1501st squadrons included . . .

. . . the C-133s (the "Weenie Wagons") assigned to the 84th ATS . . .

. . . the C-135s to the 44th ATS . . .

. . . and the C-130s from the 1513th.

The fourth type of aircraft was the C-141A, the Starlifter. On April 23, 1965, the *Golden Bear* touched down at Travis, its final home. This Travis landing marked the first assignment of any Lockheed Starlifter to an operational aircraft unit in the United States Air Force and was part of the airlift modernization begun by the Kennedy administration. This event was celebrated by the arrival of many dignitaries and the attendance of the entire population of the base. The 44th Air Transport Squadron was chosen to be the first 1501st Air Transport Wing flying squadron for the new jetlifter. Today the *Golden Bear* is part of the Travis Air Museum's aircraft collection.

Five

PEACEKEEPING MISSIONS

The history of airlift transport from Travis was linked to world crises because the base participated in almost every important deployment of American forces, advisors, or emergency-relief personnel abroad since its inception. From the widely known to the little-known—Travis was there. "Old Shakey" had been a valuable asset because it had nose doors that allowed loading without disassembly.

Travis was often involved in peacekeeping efforts that were unfamiliar to most people. In 1960 Travis began providing airlift assistance, via nine C-124 Globemasters, to deploy U.N. troops to the Congo in the international peace effort. This was known as Operation New Tape. For several years thereafter Travis C-124s and C-135s were used in troop rotations serving the Congo. Today the C-124 "Old Shakey" is part of the Travis Air Museum's aircraft collection.

The 60th Troop Carrier Wing was the direct predecessor of the 60th Military Airlift Wing, and was activated in 1948 at Kaufbeuren Air Force Base, Germany, during the famous Berlin Airlift—Operation Vittles. The airlift, between June 26, 1948, and May 12, 1949, was a response to brazen attempts by the Soviet Union to force the West out of Berlin with a blockade. Many relief supplies arrived on MATS flights. Pictured above is a C-54 flying over Germany in 1948, participating in the airlift. The Wing also served as the host unit at Rhein Main Base in the early 1950s, then was moved to Dreux Air Base in France. It played an important role in NATO's defense before it was deactivated in 1958. This wing was later reactivated as the 60th Military Airlift Wing. The Berlin crisis in 1961, which led to the construction of the Berlin Wall, resulted in President Kennedy calling upon Travis to renew American strength in Europe by deploying these aircraft. Below a C-130 loads cargo for deployment.

In 1963, a C-133 from the 1501st flew the first Gemini space capsule from Baltimore, Maryland, to Cape Canaveral, Florida, prior to its launch into space.

In 1964 Travis C-135s flew approximately 45 tons of anti-cholera serum to South Vietnam to combat an epidemic. That same year Travis also supported Operation Deep Freeze, sending a C-130E to airlift supplies to support the activities of the National Science Foundation and the Navy in Antarctica. The last C-135 left Travis on March 28, 1965, less than a month before the *Golden Bear*'s arrival. A Travis C-135 is pictured here with the record-breaking crew from the 44th Air Transport Squadron posing with brass before base farewell.

There were non-airlift activities at Travis prior to the Vietnam War. In 1961 the first GAM 77, air-to-ground, "Hound Dog" missile arrived. The B-52Gs, which had marked Travis's introduction into the jet age, had the ability to carry a pair of Hound Dogs under its wings. The partnership of the B-52G and the Hound Dog was described as SAC's "Sunday Punch." These B-52s were flown by Travis's 23rd Bombardment Squadron.

SAC's most dramatic moment of the early 1960s was the Cuban Missile Crisis of October 1962. Pictured here are airmen loading up supplies during the crisis. Like all B-52 units in SAC, the 5th Wing was alerted to the seriousness of the situation and its posture was raised to Defense Condition Two, requiring 24-hour alert and heightened readiness and training flights until the fortunate resolution by President Kennedy.

Six

VIETNAM WAR

C-97

The Travis connection to Southeast Asia actually began as early as 1954, when several MATS C-97s took part in an aeromedical airlift known as Wounded Warrior. In this operation 509 French soldiers, wounded in the decisive battle of Dien Bien Phu, were airlifted halfway around the world (14,073 miles) from Vietnam to Paris and Algeria.

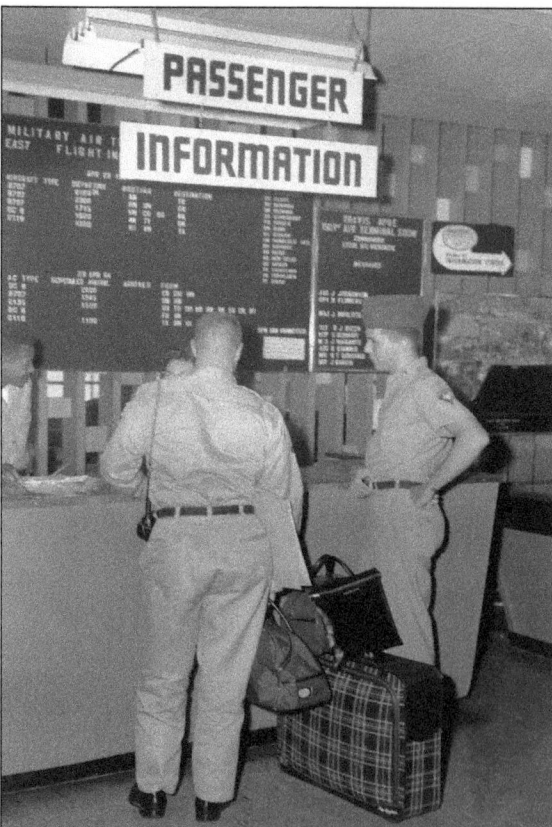

Over the next 10 years, various military transport missions passed through the base—60th Aerial Port Squadron—bound for Saigon or Vientiane, Laos. It was not until the mid 1960s, however, that the civil war raging in the area for 20 years became a direct concern to most military and civilian personnel at Travis.

Airlifts within our national borders had become the major military activity at Travis. Then in 1965 Vietnam transformed Travis into one of the most publicized passenger terminals in Air Force history—the "Gateway to the Pacific." Although the machinery that led to the Vietnam War was already in gear, Travis was now engaged in the ever-increasing tempo of its operation.

From 1954 to 1975 and beyond, the Vietnam War permeated the operations of Travis and brought on many changes. As the "Gateway to the Pacific" and the jumping-off point for thousands of American service personnel bound for the battle zone, Travis became a household word among the media, who now covered the war's progress from a totally new vantage point.

Now we're MAC! Travis personnel pose in front of a C-131 marked with the old command's name, soon to be replaced by the Military Airlift Command. During the Vietnam era, other changes also took place. On January 1, 1966, the Military Air Transport Service officially became the Military Airlift Command and a week later all MAC subordinate units were also re-designated. Today the C-131 is part of the Travis Air Museum's aircraft collection.

The expanding Vietnam War had a pervasive influence on Travis's life and every aspect of the base was affected by the war, from airlift mission assignments to increased duty times in flight hours and aerial port operations. Pictured above, bound cargo awaits transport.

In 1965, a service known as the Red Ball express became a daily mission from Travis to Saigon. This special operation was carried out by C-141, B-707, C-133, and C-130 aircraft. In its first year of operations, the express flew 695 missions, twice daily, delivering 53,994 pieces of cargo weighing 9,353 tons. The average delivery time from Travis aerial port to Saigon was just 32 hours 42 minutes, including processing and flying time.

Travis aircrews from the 60th Military Airlift Wing also participated in three major troop deployments to Vietnam between December 1965 and August 1968. The first was Operation Blue Light, which deployed 2,952 troops and 4,479 tons of equipment belonging to the Army's 25th Infantry Division. Troops are pictured here preparing for deployment from Travis boarding a C-141.

The second troop deployment from Travis to Vietnam was Operation Eagle Thrust, which sent 10,024 paratroopers and 5,357 tons of support equipment belonging to the 101st Airborne Division.

The third deployment came in July 1968 and was known as Task Force Diamond. Another 4,535 men and 480 tons of equipment were deployed from the 5th Infantry Division, Fort Carson, Colorado, to Bien Hoa to accomplish about 75 missions. All the wings of the 22nd Air Force, including the 60th, participated in this operation.

Troops are boarding a C-124 for battle in this photo. Important as they were, these super-priority operations accounted for only a small percentage of Travis's workload of transporting personnel and supplies to Southeast Asia. In fact the overwhelming majority of all personnel and cargo that were flown from the United States to Southeast Asia passed through Travis's routine airlift or supply system. Travis was the nation's busiest military port between 1965 and 1975, and the one remembered best by combatants. One serviceman reported, "When our platoon returned from Vietnam and landed at Travis, we all jumped off the plane, got down on our hands and knees and kissed the ground—we were so happy to be on U.S. soil!"

Incoming troops may have been exultant at returning to U.S. soil, but not all were so lucky. The consequences of the war in Southeast Asia were also apparent at the Travis Mortuary Affairs Office. According to its records, 10,523 military caskets from Southeast Asia passed through Travis in 1968 alone. The fallen servicemen served as a reminder to all—at Travis and everywhere—that freedom is not free.

GLOBAL RANGER

VALLEJO, CALIFORNIA

Vol. 17, No. 33   Friday, April 7, 1967

"The Global Ranger is an unofficial newspaper published weekly in the interests of personnel at Travis Air Force Base of the Military Airlift Command. It is published by the Gibson Publishing Company, a private firm, in no way connected with the Department of the Air Force. The appearance of advertisements in this publication, including inserts, does not constitute an endorsement by the Department of the Air Force of the products or service advertised."

Doolittle's Tokyo Raiders Will Visit Base Thursday

Group Celebrates 25th Anniversary

Stars of the historic 1942 B-25 bombing raid on Tokyo will rendezvous at Travis Thursday to begin a three day program commemorating the 25th Anniversary Reunion of the Doolittle Raiders. Lt. Gen. Jimmy Doolittle, now retired and living in the Los Angeles area, will officiate in presenting the Raiders - USAF Traffic Safety Awards for 1966 to representatives of the Air Training Command and Air University for outstanding achievement in the reduction of highway fatality. He will be joined at Travis by a number of surviving r a i d e r s, Bay Area dignitaries, and a restored replica of the aircraft on which they flew the famed mission, a B-25 Mitchell bomber.

Participants in the Silver Anniversary Reunion will journey to Travis Thursday morning from nearby Alameda for a view of the battle-ready B-25, complete with World War II warpaint and insignia, and a brief tour of the busy base flightline escorted by Brig. Gen. Maurice F. Casey, commander of the 60th Military Airlift Wing.

The award presentations will highlight a noon luncheon which will follow in the Officers' Open Mess. Another award presentation will also be made by Capt. Richard C. Smith, president of the Travis Junior Officers Council.

The Doolittle Raider Traffic Safety awards are presented annually as an incentive in a continuing campaign to reduce accident toll. Selection of winners in categories of large and small air commands is made by the USAF Traffic Safety Committee, including representatives of the Air Force's Directorates of Security and Law Enforcement, Transportation, Surgeon Gener-

See RAIDERS Page 2

OFF THEY GO—A B-25 Mitchell Bomber of the raiding task force of Gen. Jimmy Doolittle takes off from the carrier Hornet on its way to Tokyo April 18, 1942, while others wait their turn. Silver anniversary reunion proceedings begin Thursday morning here with award presentations and a noon luncheon highlight at the Officers' Open Mess. A restored B-25 will take to the air for a flight to Alameda Naval Air Station where it will be on display throughout Thursday, Friday and Saturday for the reunion program.

RAIDERS REUNION

From Page 1

al, Civil Engineering, Information and Aerospace Safety.

Following the award Luncheon the restored B-25 will take to the air for a flight to the Alameda Naval Air Station, where it will be on display throughout the Thursday, Friday and Saturday reunion program and will participate in a flyover. The aircraft is the property of Richmond business man Ralph Johnson, president of VANPAC Carriers, Inc., who will also attend the activities at Travis.

A Travis aircraft will deliver another historic exhibit for the reunion, a set of 80 silver goblets which are kept on permanent display at the Air Force A c a d e m y, Colorado Springs, Colo. The goblets, a gift to the Raiders from the city of Tucson, Ariz., are engraved with the names of each of the original participants, and carried in a heavy velvet-lined case.

Each yearly reunion meeting is highlighted by a formal toast of the assembled raiders to their fallen comrades. The traditional toast will be given at noon Saturday at a meeting at the Edgewater I n n, Oakland, reunion headquarters.

The Friday reunion program will include a cruise from Alameda aboard the USS Oriskany for a view of air carrier operations, a tour of the ship, a B-25 flyover and offshore aerial exhibitions of the Blue Angels Navy demonstration team.

Windup of the reunion will be a formal banquet Saturday evening at the Holiday House in Hayward with movie star Jimmy Stewart as master of ceremonies. Other luminaries in attendance will be John Charles Daly of television, Bob Considine of Hearst International News, Congressman George Miller of the 8th District, mayors and h i g h ranking Navy and Air Force officers.

In 1967 the crews of the Tokyo Raid, led by famed aviator and serviceman Jimmy Doolittle, celebrated their 25th anniversary at Travis. This raid, carried out in 1942, was the U.S. counter attack to the bombing of Pearl Harbor, which had sparked the base's construction. The B-25 crews had departed from nearby Alameda, California, aboard the U.S.S. *Hornet* to bomb Tokyo, Japan. The bombing caused Japan to change its strategy, resulting in a U.S. victory at the important battle of Midway. The Raiders' interest in Travis would culminate with their return in the year 2003 to christen an air museum.

An Ambus (ambulance bus) nestles within the petal doors of the giant C-141 Starlifter as all strive to pick up the pace to meet pressing medical needs and assist with the flow of casualties. The expanded use of 12-hour shifts in critical job areas and the assignment of additional active duty military and civilian personnel were not enough—the shortages remained critical.

SAC's B-52 is ready here for action. From 1960 until 1968, SAC operations included routine training missions of the 5th Bombardment Wing, which operated as SAC's global deterrent force. Flying B-52Gs, they were complimented by a growing number of KC-135s from the tanker fleet as support.

Pictured here are B-52s flying bombing raids. The 5th Bombardment Wing flew classified missions to Vietnam, conducting direct support missions including the following operations: Young Tiger, Cross Keys, One Buck, and Thirsty Camel. After 1968, Travis continued to support Vietnam operations and deployed KC-135s and B-52s to Southeast Asia to assist in other operations, including Bullet Shot and the famous Linebacker. Additionally, a series of B-52 raids flew over North Vietnam to help secure Hanoi's agreement to a peace settlement in Paris in 1975. The 916th Refueling Squadron commemorated these raids by dedicating "Linebacker Park" at Travis to all the personnel who played a role in these important raids.

In 1968, following the capture of the *Pueblo* by North Korea and the bloody Tet Offensive, a reserve program at Travis was implemented ahead of schedule to meet personnel shortages. The first MAC reserve associate program was officially inaugurated to assist active-duty personnel of the 63rd Military Airlift Wing. The success of this program led to its parent command, the 349th Air Mobility Wing, being added as an official associate unit at Travis in 1969. A 349th flight engineer is pictured here manning his controls aboard a C-124.

When American involvement in the Vietnam War decreased after 1970, the reserves at Travis played a role in helping homeward-bound American forces. They also assisted in the evacuation of Vietnamese refugees to the United States. (*Pacific Daily News*, Thursday April 24, 1975.)

VIETNAM PEACE PACT SIGNED; AMERICA'S LONGEST WAR HALTS
OPERATION HOMECOMING BEGINS

Lengthy negotiations in Paris between the United States and North Vietnam eventually resulted in a cease-fire agreement in January 1973. The key provisions called for a phased withdrawal of American troops and the speedy return of American prisoners of war from North Vietnam, known as Operation Homecoming.

POWs are headed back to the U.S.—three cheers, we're almost home!

The first C-141 to land at Travis with POWs was on Valentine's Day, 1973.

In the course of the next six weeks, C-141 Starlifters continued to arrive carrying additional POWs.

BRACELET RETURNED — CWO Daniel F. Maslowski receives the POW bracelet with his name engraved on it from A1C Peggy Kriesel who is stationed at Travis. Maslow-ski had noticed Airman Kriesel's sign which said "Welcome home CWO Dan Maslowski. You're my bracelet man." She had worn the bracelet for over a year when Maslowski arrived at Travis Air Force Base Thursday afternoon.

A1C returns the bracelet she wore for over a year to the POW whose name it bore. For nearly two months as 280 POWs were returned from North Vietnam, Travis's name was broadcast nearly daily in newspapers, and on radio and television.

This photograph captures a dramatic moment at Travis. This POW returns after four years in captivity.

As the POWs arrive at Travis, the press have a heyday!

Viet Orphans Arrive At Travis

eration Babylift"
t acronym for transporting
nds of refugee orphans from
am to the United States has
ne a household name around
s.

parations began long before the
light of orphans left Vietnam for
new homes in the U.S.

April 3, during a press con-
ce, President Gerald Ford

announced that MAC would assist in
transporting refugee children from
Vietnam to the United States.

The Army's Presidio in San
Francisco was one of the places set
aside to receive the refugees after
their arrival. Volunteer nurses,
doctors, and hundreds of
organizations responded to the
Presidio call for assistance.

Many charitable organizations

throughout the United States began to
notify the families that had been
waiting so long for their adopted child
to arrive.

Before long the children would be in
many homes from Florida to Maine to
California.

Preparations at Travis began with
Col. Robert Kirk being appointed
Project Coordinator for the airlift
portion at Travis.

Meetings were held and "Ope
Babylift" plans were formed ar
to work. Coordination around the
was done on a round the clock
Many offices bustled day and
preparing for the first flight.

A press center was establish
the Crosswinds Recreation Cen
accomodate the news r
representatives, such as Asso
Press and United Press Interna

(Continued on Page 14)

SMILE FOR A NEW HOME—Mrs. Rita Derman, Red Cross Volunteer with the Service to Military Families at Travis, cradles a smiling Vietnamese orphan for waiting newsmen Tuesday as she and other Travis volunteers helped offload some 286 refugee children from an ONA DC-10 jetliner. (U.S. Air Force Photo by Capt. Ned Nevels

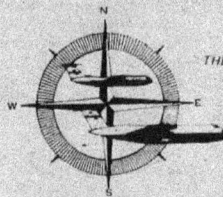

THE VOICE OF THE WORLD'S LARGEST MILITARY AIRLIFT

THE TAIL WIND

The Tailwind is an unofficial newspaper published weekly in the interest of personnel at Travis Air Force Base, Californi
Military Airlift Command. It is published by the Fairfield Daily Republic, Fairfield, California, a private firm, in no w
nected with the Department of the Air Force. Opinions expressed by publishers and writers herein are their own and are
considered an official expression by the Department of the Air Force. The appearance of advertisements, including supp
and inserts, in this publication, does not constitute an endorsement by the Department of the Air Force of products or
advertised.

| Vol. 1 | No. 45 | April 11, 1975 | Fairfield, |

0th MAW C 5 Galaxy

Despite supply efforts, the fall of Saigon was imminent in 1975. President Gerald Ford enacted measures to assist in the evacuation of the remaining American personnel in South Vietnam and locals who would be in danger due to their support of the Americans.

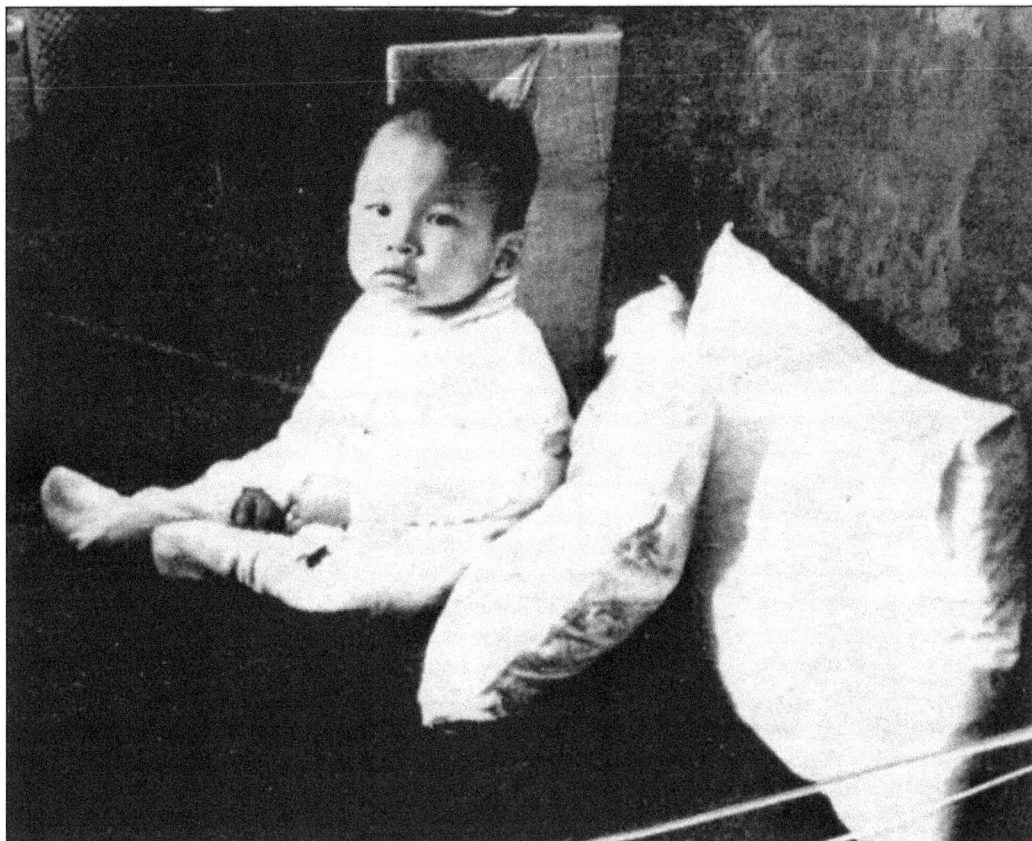

Among those given priority for evacuation was a group of orphans, both Vietnamese and racially mixed, ranging in age from a few weeks to teens. The orphan airlift to Travis was dubbed Operation Babylift. (*Pacific Daily News*, April 24, 1975.)

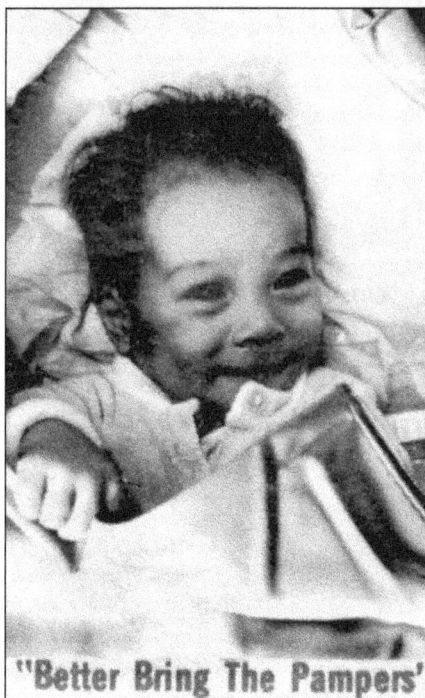

"Better Bring The Pampers'

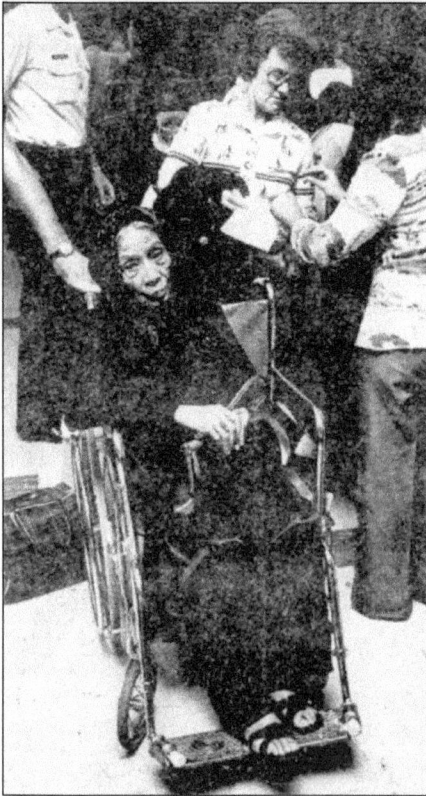

Travis aircraft have also known difficult times and unfortunately everything did not go smoothly for Babylift. On April 4, 1975, a C-5 plane crashed after takeoff in Vietnam, killing 138 people. But the heroism of the pilot saved 176. Today, a memorial stands at the Travis Air Museum in remembrance of the victims. Despite the tragic loss, the operation persevered and a total of 2,945 orphans were airlifted by 24 C-141 flights and some augmented charity flights, all directed by Travis's 22nd Air Force. All aircraft missions have potential dangers, and for those based at Travis, like this C-141A, the Golden Gate Bridge always signifies that "You're finally home."

After the war ended in 1975, Travis continued to transport refugees to the United States. The largest such transport arrived at Travis in April 1980, bringing the total number of refugees for that year to 68,394. (*Pacific Daily News*, Thursday April 24, 1975.)

Seven

SUPPLEMENTAL MISSIONS
DURING VIETNAM

The C-141 Starlifter, as shown here, is always on the move. Travis also had other missions to carry out during the war. The 60th Military Airlift Wing participated in an Embassy Run mission for six years beginning in the early 1960s. C-141 crews from Travis flew west in round-the-world flights in support of American diplomatic and military personnel.

This C-141B flies over San Francisco. Note the extended fuselage modification for air refueling, which differs from the C-141A. The C-141s of the 60th also participated in Operation Garden Plot, troop deployments to check any civil unrest that might arise due to political protest. They were meant to protect government property and officials from civil actions that got out of hand. Some assignments included carrying troops to Washington D.C. after the assassination of Dr. Martin Luther King in 1968, to Chicago during the Democratic Convention in 1968, to Washington D.C. in 1971 during the anti-war protests, and to Miami in 1972 during the Republican Convention. The military does not initiate policy, but rather follows the directives of government officials elected by the people.

Missions carried out in support of the nation's space program were more enjoyable than riot control for Travis crewmen. Beginning in 1969, Travis crews flew C-141s and C-133s to assist the Apollo 9, 10, 11 (the first lunar landing), and 12 missions. They also flew a backup mission for Apollo 11, and flew assist missions for Apollo 13, 14, and 15. Such missions extended into the 1980s with the space shuttles *Columbia* and *Challenger*. As pictured above, NASA C-141s flew to Travis for maintenance too.

Dateline 1973: Offloading an M-107 self-propelled howitzer during Operation Nickel Grass, the resupply of Israel during the Yom Kippur War.

In 1973 the 60th played a major role in Operation Nickel Grass, the airlift of American military equipment to Israel during the Yom Kippur War. C-5s and C-141s transported badly needed equipment and supplies to Tel Aviv.

The Arab oil embargo followed the Yom Kippur War. The policies of OPEC focused the world's attention on rising energy costs and dependence on Middle East oil. In 1975 MAC developed an Energy Conservation Program, having the initial goal of a 15 percent reduction in energy usage over the previous year. Pictured here are gas pumps at Travis during the 70s.

This C-5A seems happy to be at Travis! And why not? Most of the organizational changes that occurred among the squadrons at Travis between 1966 and 1975 were related to the modernization of aircraft. Almost immediately after the C-130 was transferred elsewhere and the C-124 was retired, MAC introduced the new jet transport, the C-5A Galaxy.

A Travis C-5 loads up for deployment in this shot. The appearance of the C-5 in 1970 hastened the retirement of the C-133 Cargomaster in 1971. Its retirement marked the end of propeller-driven aircraft at Travis. Since humanitarian relief missions were also part of Travis's duties, these were enhanced by the size of the Galaxy. The 60th and MAC units flew special relief missions in response to many natural or man-made disasters. These missions proved to be very rewarding to Travis personnel and included: flood assistance in Fairbanks, Alaska, in 1967, victim assistance from Typhoon Sarah on Wake Island in 1967, Hurricane Camille in Gulfport, Mississippi in 1969, quake victims in Guatemala in 1976, and 19 C-5 and 12 C-141 missions to rebuild Guam after Typhoon Pamela in 1976.

The expansion of airlift activities at Travis caused by the Vietnam War and the subsequent addition of the C-5 to the wing's inventory led to several important construction projects. The largest was the construction of the huge hangar required to house the C-5 for maintenance and repair.

78

You've come a long way, Air Force! Pictured here are women from the Army Corps before the birth of the Air Force in 1947. A later organizational change not directly connected with the wing's flying mission occurred in July 1975, when the Air Force announced the end of its dual management program for enlisted women. Until then, all Air Force women had been assigned to one unit for duty and, at the same time, attached to a Women's Air Force (WAF) squadron section for housing and counseling. After July 1975, the WAF squadron was discontinued and all women at Travis were reassigned to their regular duty squadron or headquarters squadron section.

Pictured here are civilians assisting in loading a C-5. Travis was much calmer and quieter after the hectic decade of the Vietnam War, which was characterized by frequent 12-hour shifts and saturated cargo and passenger terminals. Due to hiring freezes and downsizing, the new challenge was using a smaller work force to cope with an increased workload due to modernization and budget cuts. Reduction and conservation became the watchwords of Travis and "making do with less" became the new motto. Between 1975 and 1980, the Carter administration reduced the civilian workforce at the 60th Military Airlift Wing at Travis by 19 percent. Conservation missions became the theme for the day and in 1975 Project Pare and Project Squeeze were implemented to save energy.

Travis became a part of Fairfield following a 1966 annexation. In that year it was valued at $155 million with an annual payroll of $80 million, giving the city $400,000 in revenues that first year.

This C-141 waits while personnel ready litters for paratroopers during an operation. After 1976, the 60th Military Airlift Wing and Travis continued to be involved in many special airlift operations, joint service exercises, and humanitarian airlift missions around the world. In 1978, crews from the 60th flew special support missions to Lebanon and Zaire to assist in the peacekeeping efforts of the United Nations in those countries. In the same year Travis also began using C-5s starting with Operation Reforger.

In 1979 Travis crews also flew several missions to Alaska. These exercises continued until 1983. Pictured is a C-141 during the Alaskan exercise, Operation Jack Frost, later known as Brim Frost. This was a cold-weather exercise often held jointly with Canada.

In 1979, an Islamic revolution brought the Ayatollah Khomeini to power in Iran and the Soviet Union invaded Afghanistan. These events increased the strategic importance of the Indian Ocean and Britain's Diego Garcia Island in American military planning. Travis flew regularly scheduled missions to the island where an American Naval task force maintained port facilities. Pictured here are KC-135s, B-52s, and KC-10s lined up at Diego Garcia.

Pictured here is the Travis flight line during the 1980s with C-141s in the background. At the end of fiscal year 1982, Travis was the largest base in MAC, having property worth about $136 million and physical assets valued at more than $2.5 billion. With a combined military and civilian payroll of more than $200 million annually, the total economic impact of Travis on the area labor market in 1982 was nearly half a billion dollars, making it by far the largest civilian employer among the cities of Fairfield, Suisun City, and Vacaville.

The Travis Air Museum was established in 1983 and Lt. Gen. James H. "Jimmy" Doolittle (ret.), seen here, confers his best wishes the following year. The 1984 inscription reads, "To the Travis Air Force Museum, with every good wish—Jimmy Doolittle." The museum officially opened its doors in 1987 after acquiring the old commissary building. A new $7.1 million commissary was built in 1985. Doolittle would remain in the thoughts of museum supporters.

84

A MAC C-141 off loads supplies as Travis keeps up the pace of its ceaseless missions. Between 1983 and 1993, Travis prepared for a wide range of military contingencies including actual warfare. In July 1983, when Libya attacked Chad, the U.S. sent military aid to Chad. Travis launched one C-141 mission to Dakar, with personnel deployed to various locations in West Africa. At N'Djamena, Chad's capital, the low temperature was 115° and the work area had "no water, no food, no latrine, and no cover except underneath the wing of a parked C-47."

On October 14, 1983, communist supporters overthrew the government on the Caribbean island of Grenada and the United States invaded the island on October 25 in operation Urgent Fury. The 60th supported it with three C-5 and six C-141 Travis departures and an almost equal number of missions that were diverted from their routine scheduled routes. Pictured is a Travis C-141 in Grenada in 1983.

The main transport types used during Operation Urgent Fury were the C-130E and C-130H. The wing deployed to Pope AFB, North Carolina, with the 60th Security Police Squadron's 44-man air base ground defense flight with full combat equipment. This unit went to Grenada to help with POW transfer and later provided aircraft to Barbados, one of the staging points. American troops quickly took control of the country and on November 2, hostilities ceased, but the United States did not completely depart till June 1985.

A Travis C-141 is on its way to the Philippines. On August 21, 1983, the Philippines were in a state of virtual anarchy after the assassination of Benigno Acquino, the opposition leader. His widow was running for president against the incumbent, Ferdinand Marcos. Marcos's support waned and as Acquino declared victory, President Reagan urged Marcos to resign. A MAC C-9 immediately flew him from Clark AFB to Guam on his way to exile in Hawaii. Marcos was followed to Guam by a Travis C-141 carrying his entire entourage to Hickam AFB, Hawaii, bringing that era of Philippines history to a close.

On October 12, 1988, the David Grant Medical Center opened its doors and became the most modern hospital in the Department of the Defense. This new hospital was unquestionably the largest and most expensive construction project on the base in the 1980s.

The 60th Military Airlift Wing continued to fly the same basic aircraft, the C-5 and the C-141, although both were eventually modified. Pictured here is the flight deck of a C-141 at Travis.

Travis maintained support of existing exercises and humanitarian efforts. It also picked up new missions. Some of the most notable included transporting the Secret Service and their equipment in the United States and abroad, supporting President Reagan's trip to China in 1985, supporting President Bush's Asian trip in 1988, and his summit with Premier Gorbachev on Malta. The Travis C-5 pictured here provided presidential support.

In addition to these scheduled or routine missions, Travis flew an enormous number of individual special assignment airlift missions, including in 1984 a C-141 assignment from the 7th Military Airlift squadron. This somber mission returned the remains of the Unknown Soldier of the Vietnam War from Travis to Andrews AFB, Maryland.

Pictured here is a Travis C-141B in midair over Southern California. Travis's C-141s and C-5s also flew support missions for NASA. The C-5 supported the shuttle missions to from Vandenberg AFB in Southern California.

A base C-5 also airlifted a B-29 bomber, of World War II vintage, from China Lake to Travis, where it was restored and is currently in the Travis Air Museum's collection.

In these pictures, the C-5 has its nose raised like a knight's helmet and is in the "kneeling position." Almost every year at least one major natural or manmade catastrophe with high casualties kept Travis busy. In 1985, Travis supported victims of the Mexico City earthquake. In 1988, when a wildfire ravaged Yellowstone National Park, C-5s were flown to Camp Pendleton to airlift the Third Marine Division to fight the fire. Fighting fires caused by drought conditions was a common mission during this decade. Pictured here are crewmembers from the 349th in front of their C-5 Galaxy.

These are views of the Diego Garcia flight line along the Indian Ocean and the control tower. Politics interrupted Travis's relief efforts. War had been raging between Iraq and Iran since 1980. Fighting spilled over into neighboring areas, as Iran began to mine the Persian Gulf. This international waterway was vital to several non-combatant oil-exporting nations. On July 24, 1987, a Kuwaiti tanker, escorted by the U.S. Navy, struck a mine. The United States immediately commenced Operation Ernest Will, deploying mine sweepers and personnel to the Gulf. The 60th Military Airlift Wing supported it with four C-5s and one C-141 flying via Diego Garcia and the Indian Ocean. This war finally ended in 1988.

Operation Just Cause Panama 1989

Also in 1988 in nearby Panama, political instability peaked when President Guillermo Endara was ousted after trying to fire the head of the Panamanian Defense Forces, Gen. Manual Noriega. A general election was nullified when Noriega's candidate lost. This lead to violence and President George Bush ordered 2,000 additional troops to Panama to protect American interests. Nicknamed Operation Nimrod Dancer, the 60th launched 24 C-5s and 20 C-141 missions in support of this action.

On the American front, in 1989 when the tanker *Exxon Valdez* struck a reef in Prince William Sound, Travis quickly responded. Sending a series of C-5s to Elmendorf AFB, in Alaska, they carried landing craft, chemicals, helicopters, and cleaning equipment. Travis C-5 pilots have been trained to routinely touch down on treacherous, icy, and snow-covered runways, landing a plane almost the length of a football field. The C-5 is nicknamed, "Fat Albert" and has a unique roar as it prepares for takeoff.

Meanwhile, the situation in Panama deteriorated and after a failed coup and the killing of a Marine, President Bush announced the invasion of Panama, Operation Just Cause.

Pictured here is a Travis C-141 on the ramp during Operation Just Cause. The 60th again airlifted personnel and launched 50 C-5s and 46 C-141 missions, until January 3, 1990, when Noriega was taken into custody and the operation ended.

Eight

THE 1990S

Not long after operations in Panama ended, Operation Just Cause was dwarfed by this far greater military operation. Only two years after the war between Iraq and Iran ended in 1988, the president of Iraq, Saddam Hussein, complained that neighboring Kuwait was exceeding the permitted oil production quota and was therefore costing Iraq billions of dollars. Iraq invaded Kuwait on August 8, 1990, and annexed it. Immediately President Bush ordered an infusion of American troops and equipment to Saudi Arabia to defend it from a possible Iraqi attack.

Within six weeks, the Desert Shield airlift surpassed the magnitude of the Berlin Airlift and the supplying of Vietnam during the height of fighting in that country. It was equivalent to moving the entire cities of Davis, Lodi, or Napa, California, halfway around the world. Desert Shield became Desert Storm on January 16, 1991, as combat began.

On February 24, coalition forces led by the U.S. mounted a ground offensive, driving the Iraqis out of Kuwait, and by February 27, the Gulf War was over.

MAC had transported almost 400,000 passengers during the Gulf War (greater than the population of Oakland, California, at that time) and 362,000 tons of cargo. Between August 9 and February 28, the 60th had launched 669 C-5 and approximately 445 C-141 home station missions, and 1,132 Travis personnel had been deployed. Missions continued after the cease-fire to support residual American forces in the Gulf. It was an awesome performance for Team Travis.

The celebration for the arrival of the first Lockheed Starlifter, *Golden Bear*, at Travis in 1965, is now only a memory as retirement arrives for the oldest and most intensively used of the C-141 jetlifters. On April 23, 1990, the 25th anniversary of its delivery, *Golden Bear* was again honored as it was retired.

Two C-5s mirror each other in this photo. Only one is "camied up" (camouflaged), but both are ready for duty. In 1991, a cyclone smashed into Bangladesh, killing more than 140,000 people. Travis was there in the aftermath, providing C-5s and C-141s to airlift personnel, equipment, and supplies. Immediately following the Bangladesh catastrophe, the Philippines were shaken by the eruption of Mt. Pinatubo. Travis AFB supported the evacuation of Clark AFB, and processed thousands of evacuees. A later eruption destroyed everything in its path, including Clark AFB.

Pictured here is a Russian airlift via a C-141 during the second phase of Operation Provide Hope after the final disintegration of the Soviet Union in 1991 into 15 separate republics. These republics emerged with food and medicine shortages that prompted the United States and other countries to provide emergency aid programs. Operation Provide Hope was the task designated by the U.S., and Travis executed it beginning in 1992. Travis's C-5s airlifted supplies to Russia, Kyrgyzstan, and Armenia. In the second phase, C-5s and C-141s additionally flew to Moldavia, Ukraine, and Georgia.

Russian planes that landed at Travis for the Intermediate Nuclear Force Treaty often loaded additional supplies from Travis. Pictured here are Russian soldiers posing in front of a Travis C-141. In 1992 Travis also flew medical supplies to Mongolia following the collapse of the Communist regime. It was the first landing of an American military aircraft in that country.

Pictured here, a C-141 waits to unload relief cargo on the Horn of Africa, where after years of warfare, drought, and finally the collapse of its government, Somalia began to suffer widespread famine. By mid-1992, tens of thousands of people had died. These appalling conditions prompted an international humanitarian effort known as Provide Relief, led by the United States. On August 14, a Travis C-141 arrived with equipment to set up relief operations in Mombassa, Kenya. Then a stream of C-141s, staging from Cairo, began to land at Mombassa with aid. From there, C-130s distributed it directly to Somalia, but anarchy there prevented the aid from reaching most of the people in need. To rectify this, the U.S. offered 30,000 troops to establish security. They secured the airport in Mogadishu, the capital city. The first C-141 arrived two days later. By May 3, 1993, these aircraft had flown a total of 855 missions, collectively known as Operation Restore Hope, the largest humanitarian effort ever mounted. Travis, a full participant, also deployed several hundred personnel to the area.

In the midst of the Somalian operation, three natural disasters hit the United States in rapid succession. First, Hurricane Andrew struck Florida, moved to Louisiana, and caused more than $20 billion in damage, making it the most costly hurricane ever. Typhoon Omar battered Guam, causing the worst catastrophe in its history, and Hurricane Iniki tore across Kauai in the Hawaiian Islands, damaging one-third of its structures. In each case Travis dispatched supplies and personnel to aid the victims of these disasters via C-141s, as pictured here.

Pictured here is the 615th Air Mobility Group Operations, participating in Operation Joint Endeavor and unloading a C-5 Galaxy. From late 1993 through 1996, aircraft from Travis continued to carry out many of their assigned routine support missions and exercises. And new ones were added to the mix including Deny Flight, a follow-up imposing a no-fly zone over Bosnia for Bosnian Serb aircraft, Operation Deliberate Force, an offensive posture by the United Nations to persuade warring groups to open serious peace talks and finally Joint Endeavor, in which Travis deployed personnel and aircraft to Bosnia to guarantee peace.

The way we were: the 22nd MAS stands proudly on the Travis flight line in 1982 in front of a C-5 Galaxy. But in July of 1993, the 22nd Air Force was re-designated as a reserve unit and transferred to Dobbins AFB in Georgia. On the same date the 15th Air Force—strictly a tanker unit—moved to Travis from March AFB and immediately became the new higher headquarters of the 60th Airlift Wing.

The replacement of the 22nd Air Force by the 15th Air Force in 1993 included plans to bring tankers to Travis. Thus the KC-10 tankers (refuelers) came to Travis. Pictured here is a Travis KC-10 tanker refueling a C-141 Starlifter.

Pictured here is a KC-10 in the foreground, followed by a C-141 on the upper left and a C-5 on the upper right.

Travis continued to distinguish itself with history-making first-ever operations: in February 1994, as part of Operation Deep Freeze, a Travis C-141 landed for the first time on a "blue ice" runway—a glacier—about 20 miles from McMurdo Station in Antarctica. Every year since 1979, the 22nd Air Force (22nd Military Airlift Squadron) performed Antarctica airdrops. In 1981, it made South Pole drops as well to assist members of the National Science Foundation.

Pictured here are a C-5 and C-141 during Operation Deep Freeze.

This photo shows a C-141 preparing to dock a KC-10 for refueling. Travis's KC-10s also accomplished historic "firsts" in aviation. In 1994 they participated in multiple mid-air refueling of the new C-17, which was being tested at Edwards AFB, California, and assisted an around-the-world flight of two B-1 bombers out of Dyess AFB in Texas. This latter mission lasted 36 hours and 13 minutes and set several flight records, including the fastest time around the world non-stop.

A KC-10 demonstrates agility as it flies over the Golden Gate Bridge on San Francisco Bay.

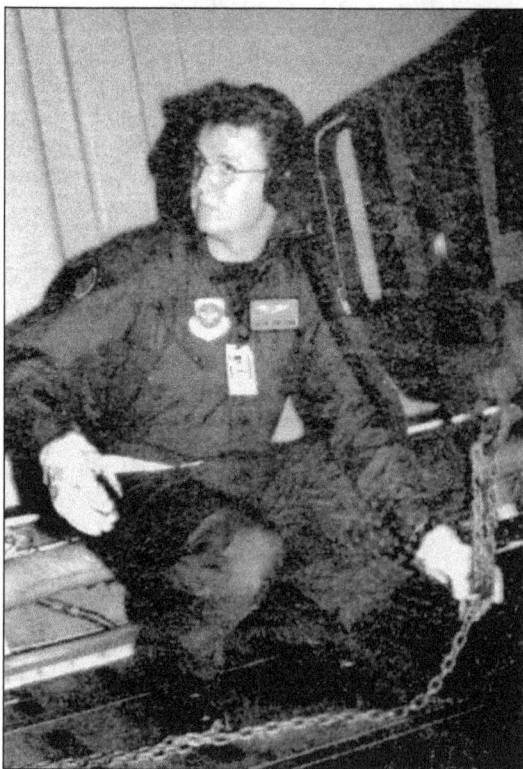

On September 19, 1994, a mass exodus of Haitian refugees approached American shores. A military junta seized power and the United States, with the support of the U.N., deployed 15,000 American troops to restore a democratic government in operation Uphold Democracy. Within ten days a legal government was restored. Travis contributed 12 C-5s and 15 C-141s as well as personnel to assist in making the Port-au-Prince airport operational again. Left, a member of the 21st Airlift Squadron participates in C-5 loadmaster training during Uphold Democracy.

Pictured here are goods prepared for an airdrop to Rwanda in 1994, with C-5 Galaxies in the background. Between July 22 and September 30, 1994, the United States launched a large-scale airlift of aid to Rwanda, which had suffered ethnic violence resulting in the slaughter of as many as half a million people. This operation was directed by the commander of the 60th Airlift Wing and was known as Operation Support Hope.

Both the active-duty personnel and the 349th Reserve Wing at Travis participated in the effort, moving almost 25,000 tons of equipment and supplies.

The majority of the equipment and supplies were moved by C-5s and C-141s that were often refueled in mid-air. In this picture a C-141 approaches a KC-10 for refueling. Also, between March 6 and June 7, 1995, members of the 60th Services Division deployed to Guantanamo Bay, Cuba, to help run a camp for refugees from Haiti.

On April 19, 1995, the bombing of a federal building in Oklahoma City shook the nation emotionally. Always ready in times of adversity at home or abroad, Travis C-141s transported a 63-member rescue team and 16.5 tons of equipment to the stricken site.

Nine

THE NEW MILLENNIUM

Pictured above is one of the many dedicated museum volunteers providing a guided tour for visitors at Travis's aviation museum. In 2000, the Travis Air Museum Foundation voted to change the name of the museum, officially opened in the old commissary building in 1987, to the Jimmy Doolittle Air and Space Museum in honor of this great aviator. The foundation also sought to build a new museum with private funding. Doolittle participated in the 1942 counter attack on Japan after Pearl Harbor (see page 60). Shown at right is the new museum patch.

On September 11, 2001, the world was stunned when terrorists crashed jetliners into the World Trade Center in New York City, the Pennsylvanian countryside, and the Pentagon. Travis quickly responded to the attacks, and soon received a visit by President George W. Bush, Military Commander-in-Chief, who addressed a Travis AFB crowd on October 17, 2001. He reiterated that America "will not fail" against the war on terrorism. (USAF photograph by Nan Wylie.)

In this picture, the surviving Jimmy Doolittle "Tokyo" Raiders pose in front of a coveted B-25 at the Nut Tree Airport in Vacaville as they celebrate their 61st reunion. The Doolittle Raiders honored Travis with their reunion in 2003, just as they had in 1967. This time the base shared the privilege of hosting these heroes with the surrounding cities of Fairfield, Suisun City, and Vacaville. Worldwide, hundreds of volunteers graciously donated their time and services for this tribute to commemorate their extraordinary service to their country. The Doolittle Raid was not only a triumph for the Army/Air Force, but the Navy as well. The success of the reunion was overwhelming as thousands attended. Despite Jimmy's passing in 1993, his legacy and spirit live on in his namesake museum and in thoughts and hearts throughout the world. (USAF photograph by Kristina Cilia.)

The new, still evolving C-17, the Globemaster III, seen here, will soon replace the Air Force's aging fleet of C-5s and C-141s. Travis has been selected to make this transition into the new era.

As Travis approached the 21st century, its role of providing global reach for the United States is now stronger than ever.

One of the largest operational Air Force bases in the nation and the most important on the West Coast, Travis AFB has been vital to America's security and the strengthening and promotion of her interests. (USAF photograph below by Scott Dreier.)

This book serves to recount several Travis AFB stories, but could not possibly include all the successes and tasks because the numbers are overwhelming.

Travis personnel have been at every crucial turn in America's history since 1943, supporting and strengthening her democratic legacy at the risk of life and limb.

Travis AFB and its personnel have quietly gone about their duty for more than half a century, the unsung heroes often overlooked by most Americans, helping the nation during many adversities.

Travis's mission, as part of the United States military, is to protect the shores of America and aid other lands throughout the world.

They maintain a constant, steadfast watch, keeping Americans out of harm's way, garnering successes that never made the front page.

And by raising one hand to swear an oath to defend our nation, Team Travis was there.

Now we know.

Thank you, Team Travis.

We salute you!

Visit us at
arcadiapublishing.com